About the Marine Sanctuaries Conservation Series

The National Oceanic and Atmospheric Administration's National Ocean Service (NOS) administers the National Marine Sanctuary Program (NMSP). Its mission is to identify, designate, protect and manage the ecological, recreational, research, educational, historical, and aesthetic resources and qualities of nationally significant coastal and marine areas. The existing marine sanctuaries differ widely in their natural and historical resources and include nearshore and open ocean areas ranging in size from less than one to over 5,000 square miles. Protected habitats include rocky coasts, kelp forests, coral reefs, sea grass beds, estuarine habitats, hard and soft bottom habitats, segments of whale migration routes, and shipwrecks.

Because of considerable differences in settings, resources, and threats, each marine sanctuary has a tailored management plan. Conservation, education, research, monitoring and enforcement programs vary accordingly. The integration of these programs is fundamental to marine protected area management. The Marine Sanctuaries Conservation Series reflects and supports this integration by providing a forum for publication and discussion of the complex issues currently facing the National Marine Sanctuary Program. Topics of published reports vary substantially and may include descriptions of educational programs, discussions on resource management issues, and results of scientific research and monitoring projects. The series facilitates integration of natural sciences, socioeconomic and cultural sciences, education, and policy development to accomplish the diverse needs of NOAA's resource protection mandate.

Comments on Hydrographic and Topographic LIDAR Acquisition and Merging with Multibeam Sounding Data Acquired in the Olympic Coast National Marine Sanctuary

Steven S. Intelmann

Olympic Coast National Marine Sanctuary, NOAA

U.S. Department of Commerce
Carlos M. Gutierrez, Secretary

National Oceanic and Atmospheric Administration
VADM Conrad C. Lautenbacher, Jr. (USN-ret.)
Under Secretary of Commerce for Oceans and Atmosphere

National Ocean Service
John H. Dunnigan, Assistant Administrator

National Marine Sanctuary Program
Daniel J. Basta, Director

Silver Spring, Maryland
August 2006

DISCLAIMER

Report content does not necessarily reflect the views and policies of the National Marine Sanctuary Program or the National Oceanic and Atmospheric Administration, nor does the mention of trade names or commercial products constitute endorsement or recommendation for use.

REPORT AVAILABILITY

Electronic copies of this report may be downloaded from the National Marine Sanctuary Program web site at *www.sanctuaries.nos.noaa.gov*. Hard copies may be available from the following address:

National Oceanic and Atmospheric Administration
Office of National Marine Sanctuaries
SSMC4, N/ORM62
1305 East-West Highway
Silver Spring, MD 20910

COVER

Perspective view of Cape Flattery and Tatoosh Island. Hydrographic and topographic LIDAR data merged with shallow water multibeam soundings and gridded at 4 meter resolution. The nautical chart was masked in Fledermaus with data obtained from both the 10 kHz and 1 Khz lasers.

SUGGESTED CITATION

Intelmann, S.S. 2006. Comments on hydrographic and topographic LIDAR acquisition and merging with multibeam sounding data acquired in the Olympic Coast National Marine Sanctuary. Marine Sanctuaries Conservation Series ONMS-06-05. U.S. Department of Commerce, National Oceanic and Atmospheric Administration, National Marine Sanctuary Program, Silver Spring, MD. 18 pp.

CONTACT

Steven S. Intelmann
Habitat Mapping Specialist
NOAA/National Marine Sanctuary Program
N/ORM 6X26
115 E. Railroad Avenue, Suite 301
Port Angeles, WA 98362
(360) 457-6622 X22
steve.intelmann@noaa.gov

ABSTRACT

In April 2005, a SHOALS 1000T LIDAR system was used as an efficient alternative for safely acquiring data to describe the existing conditions of nearshore bathymetry and the intertidal zone over an approximately 40.7 km^2 (11.8 nm^2) portion of hazardous coastline within the Olympic Coast National Marine Sanctuary (OCNMS). Data were logged from 1,593 km (860 nm) of track lines in just over 21 hours of flight time. Several islands and offshore rocks were also surveyed, and over 24,000 geo-referenced digital still photos were captured to assist with data cleaning and QA/QC. The 1 kHz bathymetry laser obtained a maximum water depth of 22.2 meters. Floating kelp beds, breaking surf lines and turbid water were all challenges to the survey. Although sea state was favorable for this time of the year, recent heavy rainfall and a persistent low-lying layer of fog reduced acquisition productivity. The existence of a completed VDatum model covering this same geographic region permitted the LIDAR data to be vertically transformed and merged with existing shallow water multibeam data and referenced to the mean lower low water (MLLW) tidal datum. Analysis of a multibeam bathymetry-LIDAR difference surface containing over 44,000 samples indicated surface deviations from −24.3 to 8.48 meters, with a mean difference of −0.967 meters, and standard deviation of 1.762 meters. Errors in data cleaning and false detections due to interference from surf, kelp, and turbidity likely account for the larger surface separations, while the remaining general surface difference trend could partially be attributed to a more dense data set, and shoal-biased cleaning, binning and gridding associated with the multibeam data for maintaining conservative least depths important for charting dangers to navigation.

KEY WORDS

Hydrographic LIDAR, Topographic LIDAR, SHOALS, multibeam, VDatum, Olympic Coast National Marine Sanctuary

TABLE OF CONTENTS

LIST OF FIGURES AND TABLES

INTRODUCTION

Airborne LIDAR (light detection and ranging) technology developed throughout the 1990s is a new tool for use in bathymetric mapping. Today, airborne LIDAR bathymetry (ALB) has fully developed into a mature technology that is a cost effective means for quickly and efficiently obtaining bathymetry information, even in areas of hazardous conditions or in remote locations too difficult to survey through conventional acoustic methods (MacDonald 2005).

The initial commercialization of this new technology is, in part, traceable to a 1998 Memorandum of Agreement (MOA) between the U.S. Army Corps of Engineers (USACE) and the Naval Meteorology and Oceanography Command, which established the Joint Airborne LIDAR Bathymetry Technical Center of Expertise (JALBTCX) in an effort to create a mechanism for helping shape future developments and capabilities of ALB mapping (MOA1998). This original MOA was superceded in 2002 (MOA2002) by creation of a new MOA that added the National Oceanic and Atmospheric Administration's (NOAA) National Ocean Service (NOS) and Office of Marine and Aviation Operations (OMAO) as partners. Expanding the breadth of partnership to include various facets of NOAA harnessed other existing capabilities, survey knowledge, tidal determination capabilities, and other unavailable resources.

In March 2005, under the NOS agreement code, a Support Agreement (MOA-2002-047 SA #001/1223) was written to facilitate the acquisition of LIDAR data for NOAA's Olympic Coast National Marine Sanctuary (OCNMS) in Washington State. Under this Agreement, Fugro Pelagos, Inc. (FPI) was contracted by GRW Engineers to conduct a site survey for the USACE (and ultimately OCNMS) along a portion of coastline from Koitlah Point to Cape Alava within the OCNMS. OCNMS' objectives for the survey effort were three-fold: 1) to obtain existing conditions of the nearshore bathymetry and intertidal zone along a select portion of shoreline that is not ascertainable through ship-based acoustic data acquisition techniques due to hazardous surf conditions, 2) to assess the performance of ALB technology along both exposed and unexposed stretches of coastline, and 3) to vertically transform the ALB data and assess agreement with existing shallow water multibeam bathymetry data collected in the same area.

SURVEY AREA

Approximately 40.7 km^2 (11.8 nm^2) of nearshore bathymetry and coastline between Koitlah Point and Cape Alava, in the general vicinity of Cape Flattery, and bounded by coordinates 48° 08'38'' N, 124° 47'34''W, and 48° 24'46'' N, 124°38'11''W (Figure 1) were proposed for surveying with a SHOALS 1000T LIDAR system. Survey flights occurred, or were attempted, between April 19 and April 24, 2005, with data being logged from 860 nm (1,593 km) of track lines in just over 21 hours of flight time. Bathymetry data was targeted between the shoreline and the approximate 15 meter bathymetry contour or laser extinction, which ever came first. Bathymetry data from the 1 kHz laser was acquired at 400 meters altitude, at 125 knots and with coverage plan designed to obtain 4 by 4 meter spot spacing using 25 percent line overlap. During project design, topographic survey lines were created to obtain heights 100 meters shoreward or to the mean higher high water (MHHW) line, which ever came first. Data from the 10 kHz topographic laser was acquired at 700 meters altitude, at 155 knots and with coverage designed to obtain 2 by 1.6 meter spot spacing. Several islands and offshore rocks were also surveyed, and over 24,000 geo-referenced digital still photos were simultaneously captured to assist with data cleaning and QA/QC.

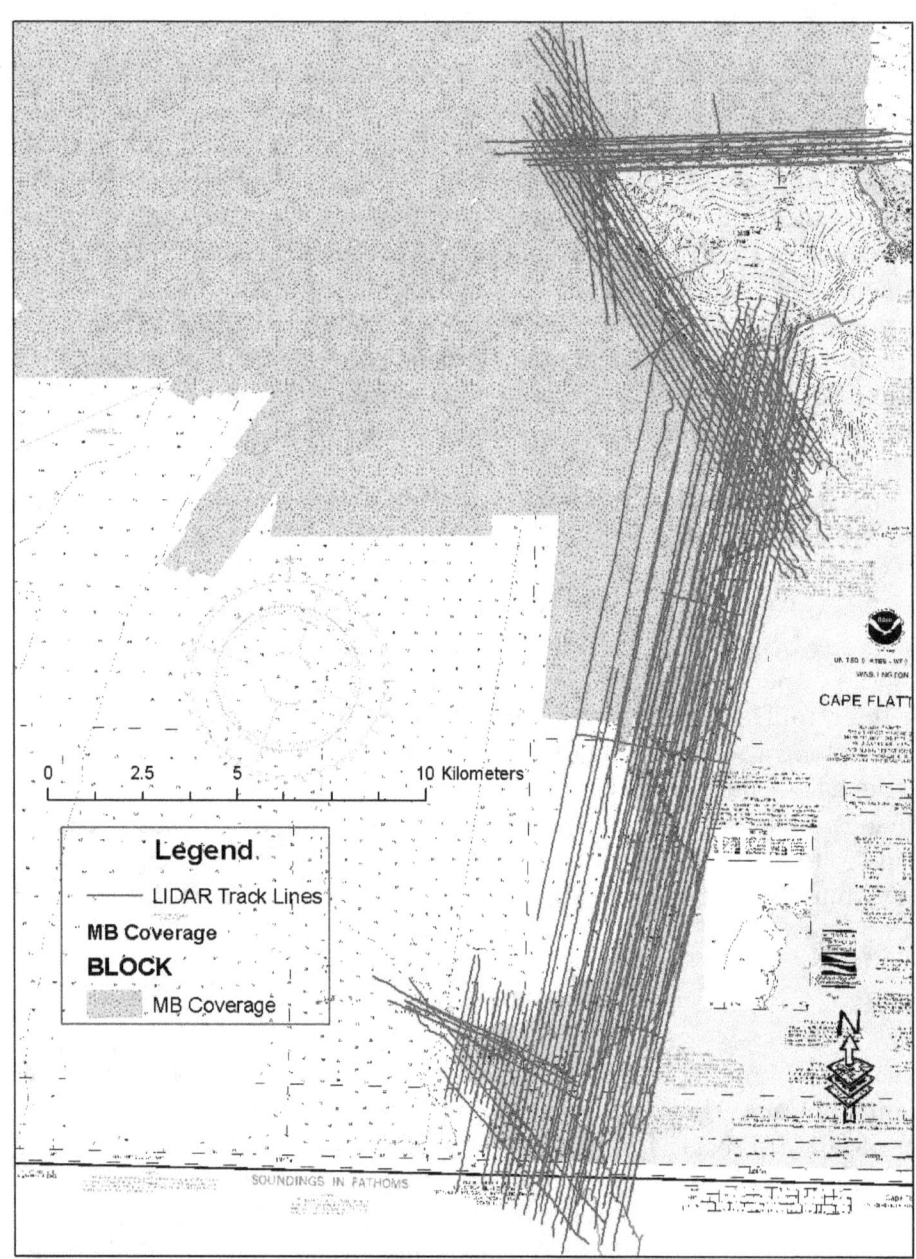

Figure 1. Cape Flattery LIDAR flight track lines, Koitlah Point to Cape Alava, shown with existing area of multibeam sounding data (blue polygon).

BASIC LIDAR FUNCTION

The basic principle of topographic LIDAR operation involves transmitting light in the form of a laser onto a mirror that is rotated at a high rate of speed. The rotating mirror projects the laser as a series of pulses onto the ground. Light is reflected back to the instrument, and with known vessel position, two-way time travel is used to ultimately compute the positional measurement of reflecting objects. ALB systems, however, are more sophisticated in theory because they must compensate for multiple reflective surfaces (i.e., the water surface and the seafloor). As such, ALB systems function as a multiple phase system where two different waves (infrared and green) are respectively used to detect the sea surface and sea floor. The newest generation of ALB systems, such as the one used for this particular survey, employs both types of LIDAR systems making them ideal for mapping shallow water intertidal zones as both height and sounding measurements can be delivered as a seamless data set. A seamless hydrographic and topographic LIDAR data set can then potentially be merged with acoustically derived single or multibeam bathymetry data, although a vertical datum transformation will likely be needed if the LIDAR data is acquired in an orthometric vertical datum and the multibeam data is referenced to some tidal datum such as MLLW (National Research Council 2004).

LIDAR DATA ACQUISITION AND PROCESSING

Details of the LIDAR acquisition and processing are in the attached Appendix, entitled "Hydrographic & Topographic LIDAR Acquisition, Northwest Coast, Washington, Neah Bay to Cape Alava, WA Survey Report" prepared by the FPI data center.

LIDAR SURVEY PRODUCT

Numerous shoals and hazardous surf conditions exist throughout the survey area, thereby precluding the use of ship-based acoustic multibeam for mapping the extreme nearshore zone in this area. LIDAR provided an efficient alternative for safely acquiring additional information to describe existing conditions of the nearshore bathymetry and intertidal areas throughout the Koitlah Point to Cape Alava region. As a general rule, the SHOALS 1000T (see Appendix for full description) is designed to obtain bottom depths of roughly 2.5 to 3 times the secchi depth. On April 15, 2004, 11 secchi measurements were taken throughout the proposed survey area in an effort to anticipate potential LIDAR performance (Figure 2). Secchi measurements ranged from 3.2 to 8.7 meters, indicating that bottom detection could potentially be achieved to anywhere between 8 and 26 meters, depending on water clarity at the time of the survey.

All topographic survey lines were fully completed and achieved with desired results; however, consistently poor weather conditions and a low-lying fog layer prevented any data acquisition from occurring on two of the six available survey days compromising bathymetry data acquisition efforts. Temporary cloud cover further restricted flying time on three additional days. As a result, significantly less data was acquired during the six-day time frame than had been hoped for. Very little bathymetry data was obtained in the southern region of the survey area due to weather constraints on survey time, poor water clarity resulting from recent significant rainfall and dense areas of kelp in the water column (Figure 3).

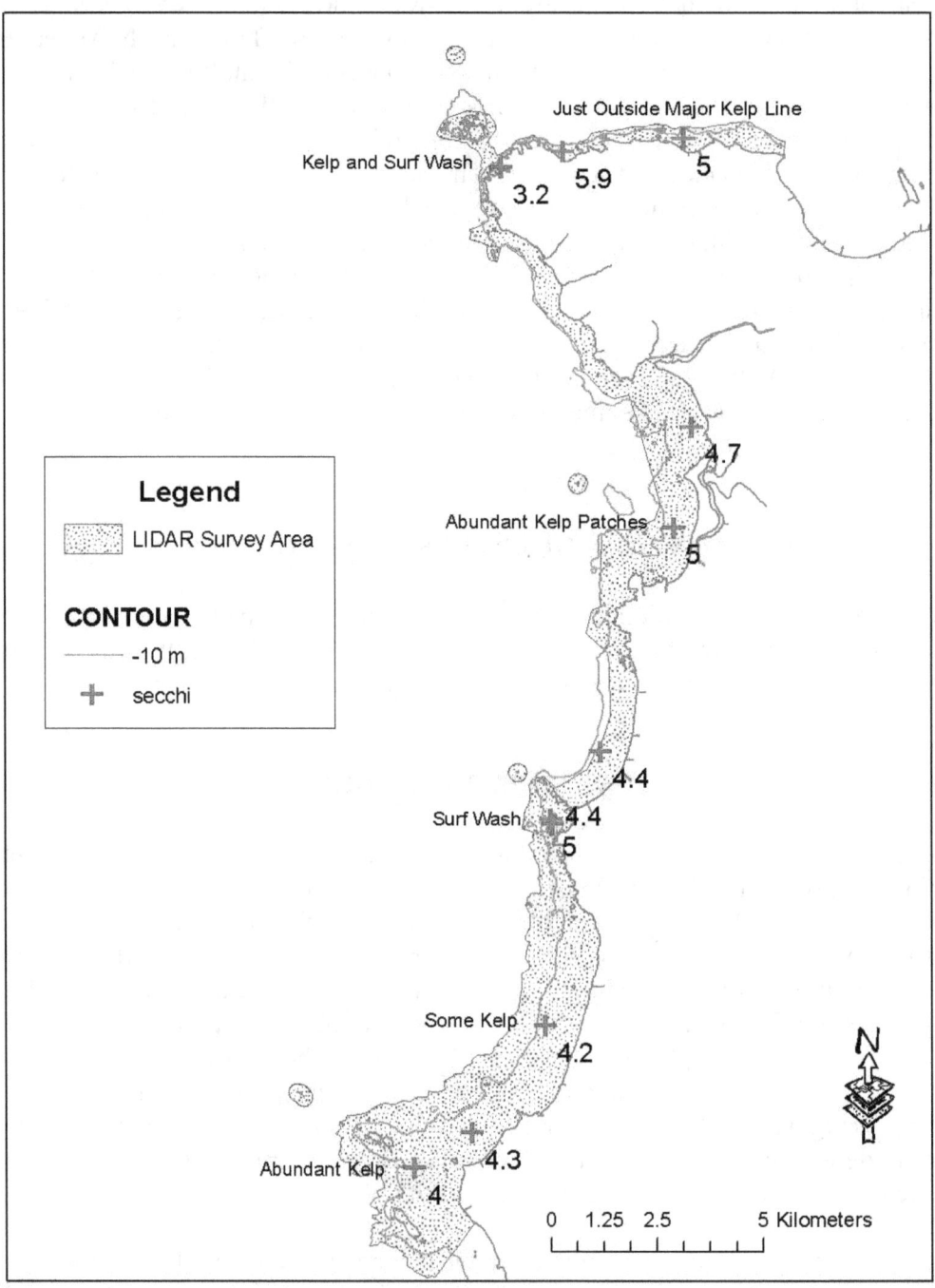

Figure 2. Locations of secchi measurements taken April 15, 2004. Secchi readings are in meters. Visual observations, if noteworthy, were also recorded at each site at the time of the measurement. The blue line represents the 10 m contour and purple polygon depicts the proposed survey boundary.

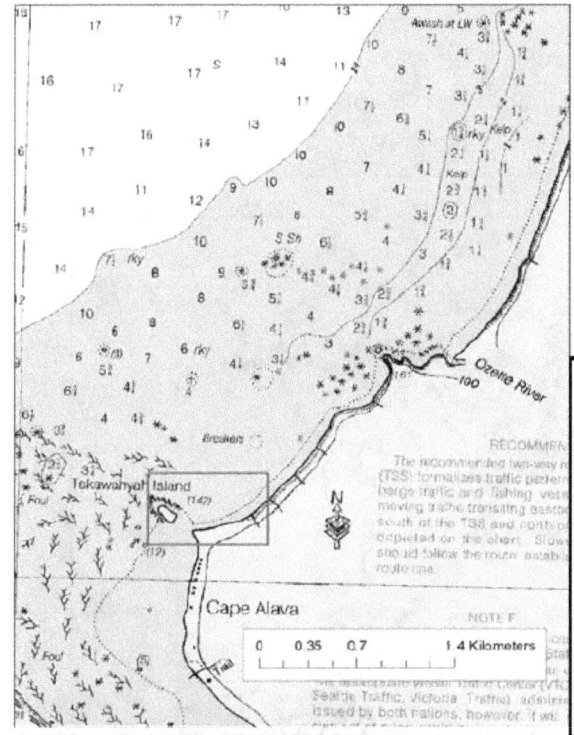

Figure 3. The graphic to the left represents the inset region of aerial photography provided below.

Surf breaks, kelp beds and turbid water all created challenges for acquiring bathymetry data in the southern portion of the survey area. All three challenges are clearly visible in the digital photo mosaic below. Images are overlaid with the accepted height data acquired from both the 1 kHz (bathymetry) and 10 kHz (topography) lasers. Example data are from the Tskawanyah Island area, and are gridded at 4 m resolution.

The effect of water clarity and inclement weather on survey productivity is immediately apparent when overlaying the actual LIDAR coverage achieved with the proposed survey area as shown in Figure 4.

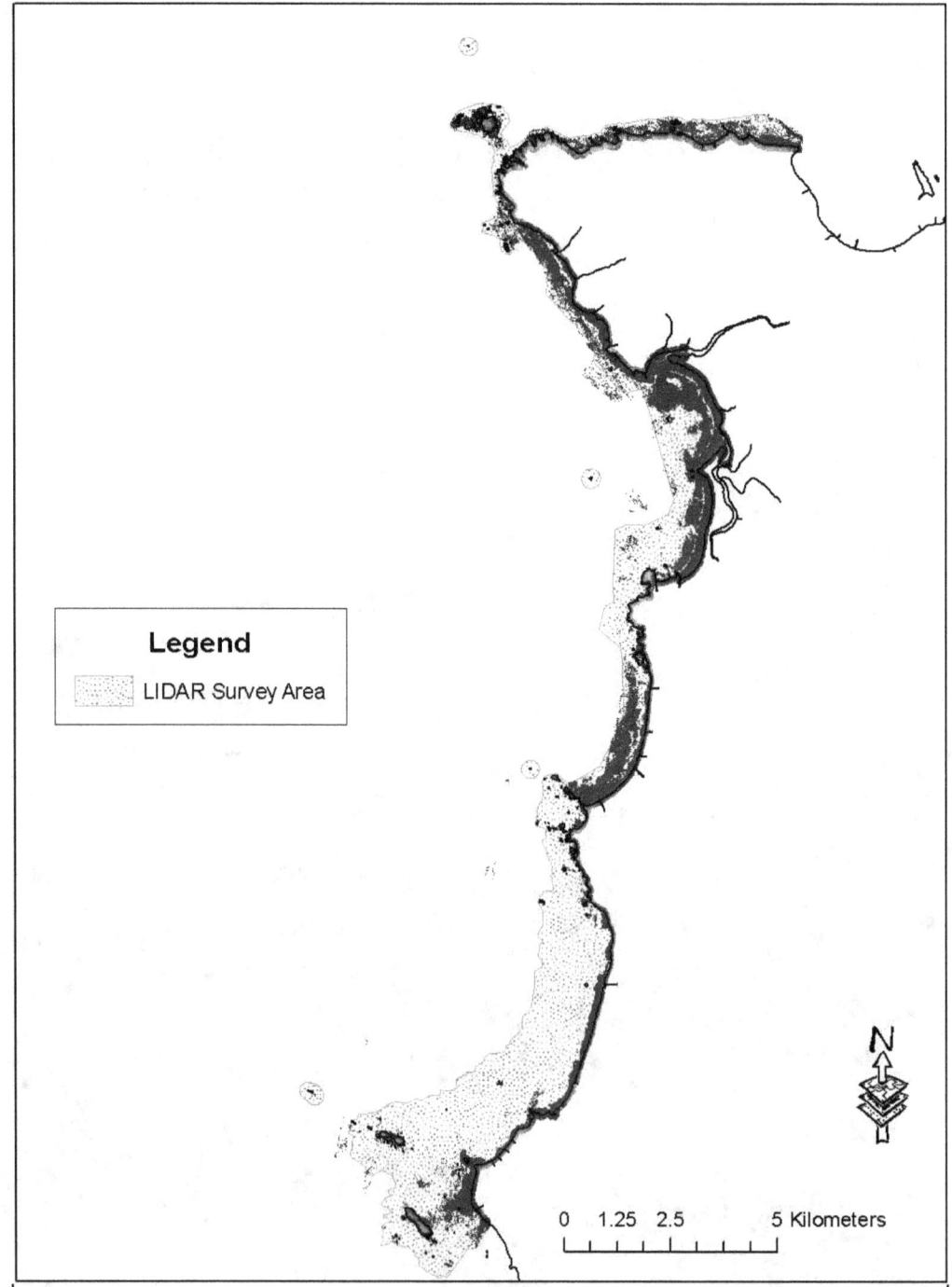

Figure 4. Combined hydrographic and topographic LIDAR coverage achieved over the six-day survey period. Heights and soundings are color ramped from red (highest surface elevation) to purple (deepest water depth). Grey polygon represents the proposed coverage area.

The maximum water depth obtained by the 1 kHz bathymetry laser was 22.2 meters, and it was located along the western shoreline of Tatoosh Island. This was near the maximum range that the secchi readings, although collected the previous year, suggested could be achieved.

DATA QUALITY ALONG EXPOSED AND UNEXPOSED COASTLINE

During the planning phase, there was concern over the suitability of LIDAR for being used as a bathymetry acquisition tool along this particular stretch of coastline. The degree of exposure to open ocean conditions and prevailing swell direction can at times produce considerable surf breaks in this area, through which the bathymetry laser would not penetrate. Having the survey designed around Cape Flattery, it provided an opportunity to assess the impact of open ocean exposure on the ability to successfully acquire ALB data in this area, as 6 km of relatively unexposed coastline exist immediately adjacent to roughly the same length of exposed coastline along the western edge of Cape Flattery (Figure 5).

Visual examination of LIDAR coverage did not suggest better LIDAR performance along the exposed stretch of coastline as compared to that along the unexposed (Figure 5). In fact when gridded at a 4 meter resolution, bathymetry data was obtained throughout 1.18 km^2 of the proposed 3.5 km^2 of exposed coastline (33.7 percent), in comparison to 1.03 km^2 of the proposed 3.05 km^2 (33.7 percent) of unexposed coastline along the Cape Flattery portion of the survey area. Furthermore, the deepest soundings obtained in the entire survey area (> 20m depth) were acquired along the exposed side of Tatoosh Island.

The bathymetry data along both these particular stretches of coastline were acquired on April 19, 22 and 23, 2006. Archived wave statistics obtained from the NOAA National Data Buoy Center (NDBC) for Station 46087, located approximately 11 km north of Tatoosh Island (48°29'38" N, 124°43'38" W), indicated that relatively calm seas occurred in this area throughout the entire survey effort. The buoy data shows the general wave trend for this particular time frame consisted of swell heights being less than 1.75 meters on a greater than 10 second swell period and with wind waves being less than 0.75 meters spaced on an roughly 4 second wave period (Figure 6). In general, both swell and wind direction came directly from the west during the survey flights (Table 1). The degree of swell and wave conditions at the time of survey were mild enough to not negatively impact the ALB data along the exposed coastline any more than that along the unexposed coastline. This indicates that other factors such as increased turbidity from recent rainfall and floating kelp played a more significant role than exposure in impairing laser penetration along this particular stretch of coastline.

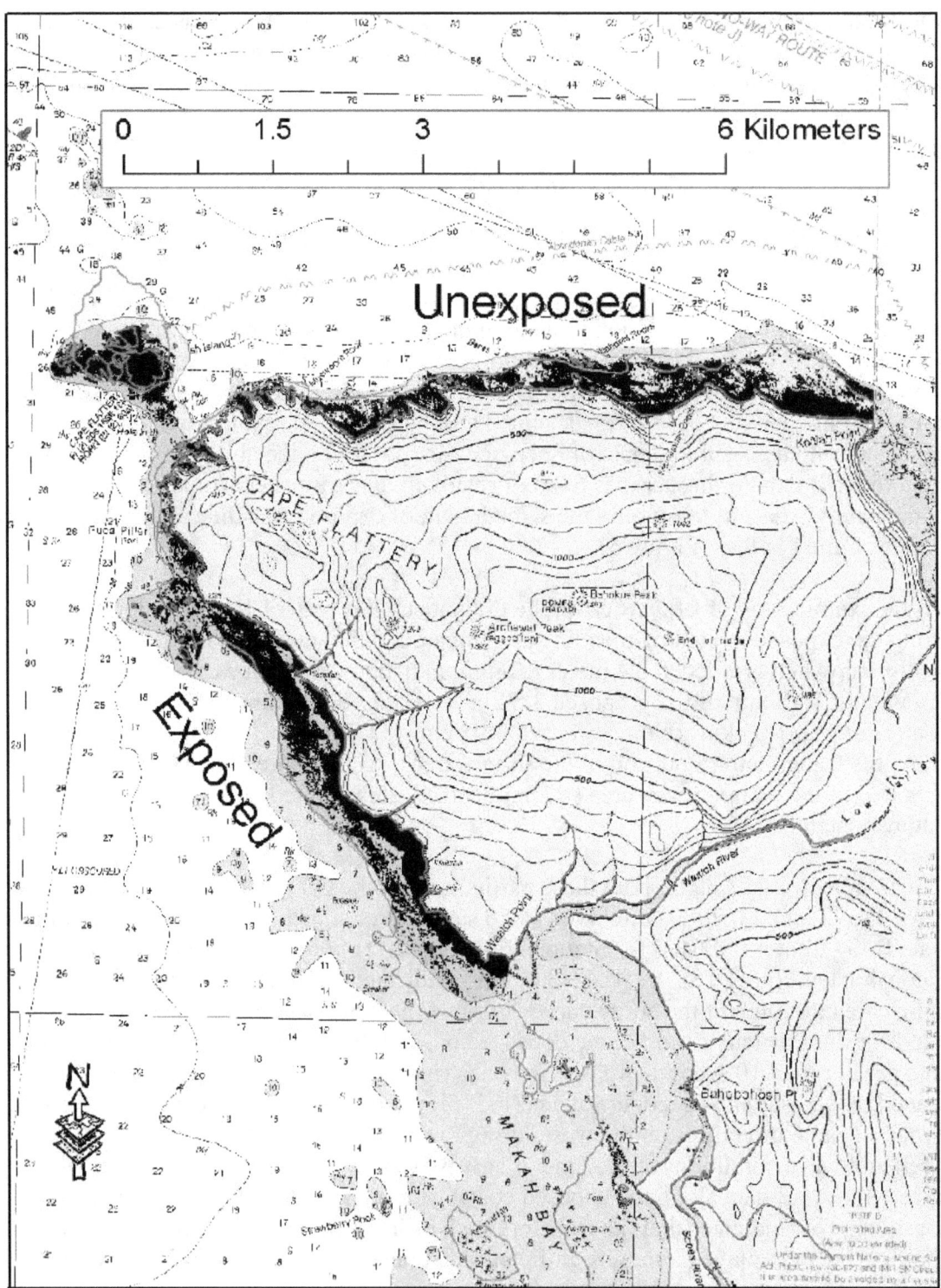

Figure 5. Adjacent stretches of exposed (purple) and unexposed (green) coastline within the proposed survey area. Achieved LIDAR bathymetry data are shown (black), with 10 meter bathymetry contour (blue).

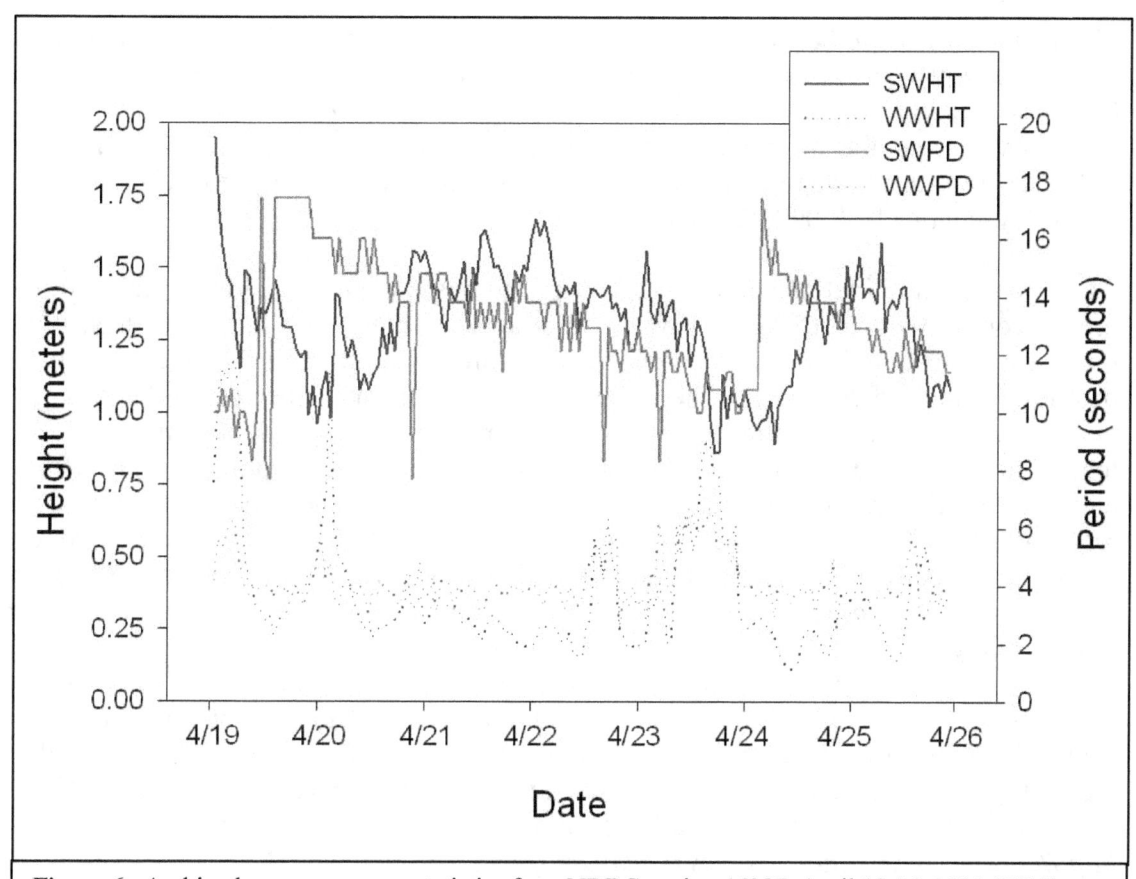

Figure 6. Archived wave summary statistics from NDBC station 46087, April 19-26, 2005 (UTC). SWHT = swell height, WWHT = wind wave height, SWPD = swell period, WWPD = wind wave period.

Table 1. Archived wave summary statistics from NDBC station 46087, April 19-26, 2005 (UTC). SWDIR = swell direction, WWDIR = wind wave direction. Values are azimuth degrees.

	SWDIR	WWDIR
Mean	272	265
Min	252	211
Max	294	357

LIDAR AND MULTIBEAM DATA MERGE

Through a partnership between OCNMS, NOAA's OCS and OMAO, high resolution bathymetry (HRB) was collected on various opportunistic occasions during the months of October from 2001 to 2004 aboard the NOAA ship *RAINIER* (Intelmann et al. 2006). Shallow water multibeam sounding data were cleaned according to NOAA standards (NOAA 2003), and were referenced to the MLLW tidal datum using the tide gauge at Neah Bay (Station 9443090) for datum control.

DGPS in the ellipsoidal datum of NAD83 supplied both the positional information and project control for the LIDAR survey. In order to easily merge the data set with existing multibeam sounding data, survey instructions required the data to be projected during post-processing and delivered in the UTM Zone 10 projection. The Geoid99 height model was used to convert the vertical datum from the ellipsoidal 3-D datum of NAD83 to the orthometric vertical datum NAVD88. In order to accurately merge the LIDAR data with existing multibeam sounding data, the data sets must be in the same vertical reference frame (Milbert 2002). Since the multibeam sounding data were referenced to an averaged tidally-derived vertical datum (MLLW) and the LIDAR data were referenced to an orthometric vertical datum (NAVD88) based on Mean Sea Level (MSL), a VDatum model (Spargo et. al 2006) was needed to vertically transform the LIDAR data to MLLW for compatibility and comparison with the multibeam sounding data. The VDatum model relates the NAVD88 to MLLW by using a grid or zone of tide model comparisons (Figure 7) with known leveled tide benchmark stations to better account for the spatial variability of tidal dynamics over a given area (Milbert 2002; Spargo et. al 2006). The VDatum model that was used is available for download at *http://chartmaker.ncd.noaa.gov/csdl/vdatum_projectsWA.htm.*

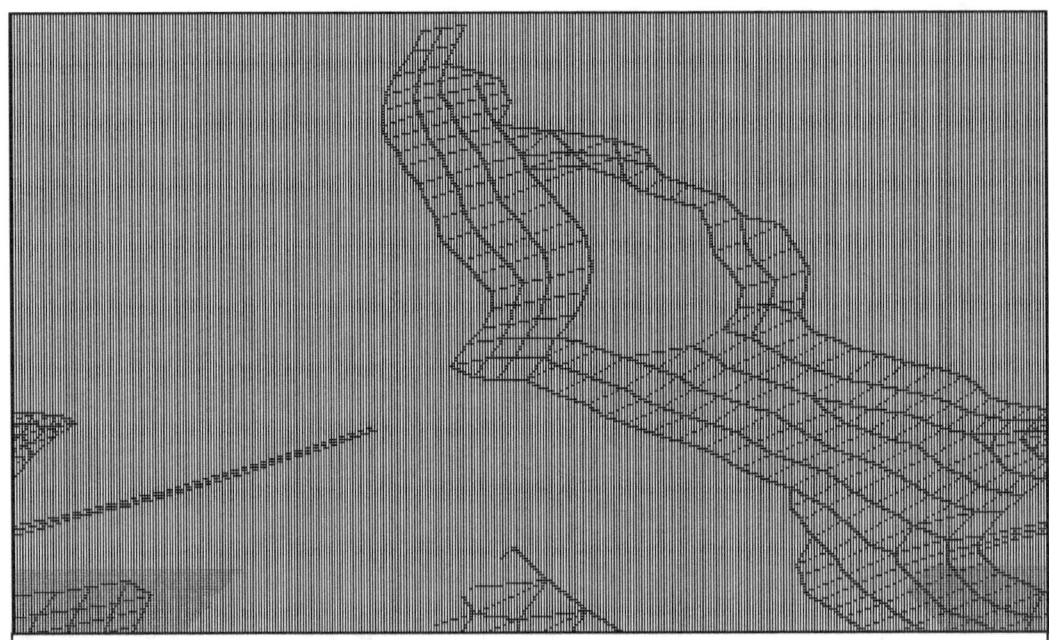

Figure 7. VDatum tide modeling grid for the San Juan Islands, Strait of Juan de Fuca and Puget Sound (Spargo et al. 2006).

Both the multibeam sounding and LIDAR xyz data were imported into Fledermaus using AVGGrid with 4 meter grid spacing (Figure 8), and then exported as separate ESRI floating-point ASCII grids for surface comparison.

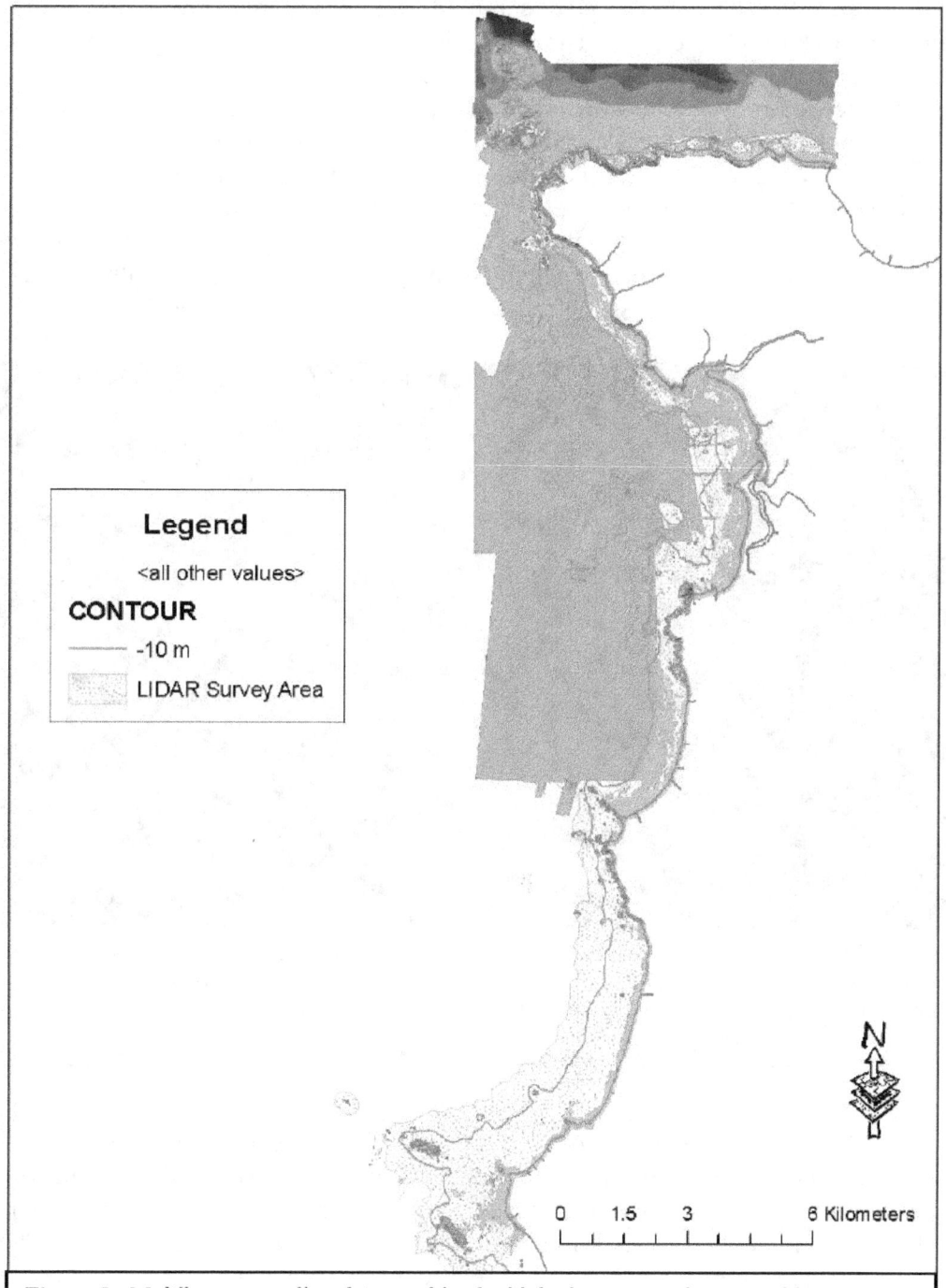

Figure 8. Multibeam sounding data combined with bathymetry and topographic LIDAR data sets. Data are gridded at 4 meter resolution. Visible areas of the brown-dotted polygon are "holidays" in LIDAR acquisition due to poor water clarity. 10 m isobath is shown as the blue line.

The two ASCII grids were converted to polygon features and intersected in ARCInfo to create a mask polygon. ARCInfo grid was then used to mask each of the ASCII grids with the intersect polygon, thus creating two new grids that were only populated with cells for which each data set had in common. The two new grids, which now contained only data common to both surfaces, were exported again as separate ASCII grid files and then imported back into Fledermaus using AVGGrid with 4 meter spacing. A difference DTM was created in Fledermaus by subtracting the LIDAR surface from the multibeam sounding surface to produce a new difference surface. The difference surface contained over 44,000 samples for which each grid had in common, with depth differences ranging from –24.3 to 8.48 meters, a mean difference of –0.967 and standard deviation of 1.762 meters (Figure 9). Depth differences between the LIDAR and multibeam surfaces were further compared to the acoustic multibeam data alone, assuming the acoustic soundings represent a more accurate reference benchmark surface (Riley 1995). As with Riley (1995), no attempt was made to assess individual error contribution to each data source.

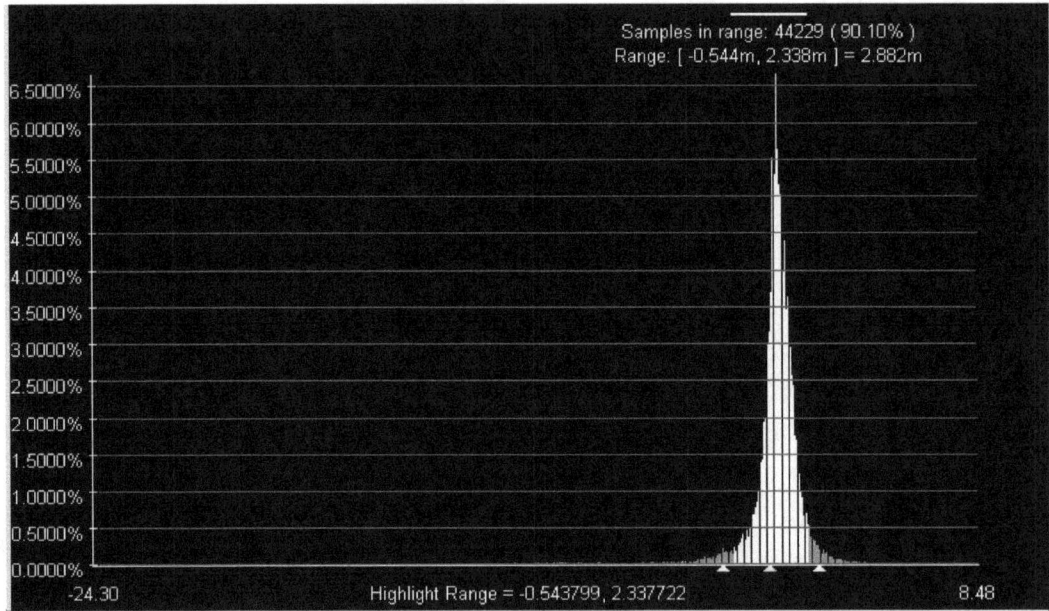

Figure 9. Depth difference histogram. The 2 * Std. Deviation (90 percentile) range is highlighted in yellow. Depth difference in meters (Multibeam Surface – LIDAR Surface) is represented by the X-axis, and the Y-axis is the percentage of samples contained in each difference bucket category.

The extreme deviations observed in the least squares fit between the multibeam and LIDAR surfaces are attributed to false bottom detections in the LIDAR depths due to water column interference (Figure 10). When examining only the 90th percentile sample range (-0.54–2.33 m depth difference), the multibeam sounding data are for the majority about 1 to 1.5 meters more shoal than the LIDAR data. This is not surprising since multibeam systems produce more dense data thus leading to delineation of more shoal features. This shoal bias of the multibeam data can be more easily visualized through Figure 11 where the enhanced surface difference is readily distinguished by examining

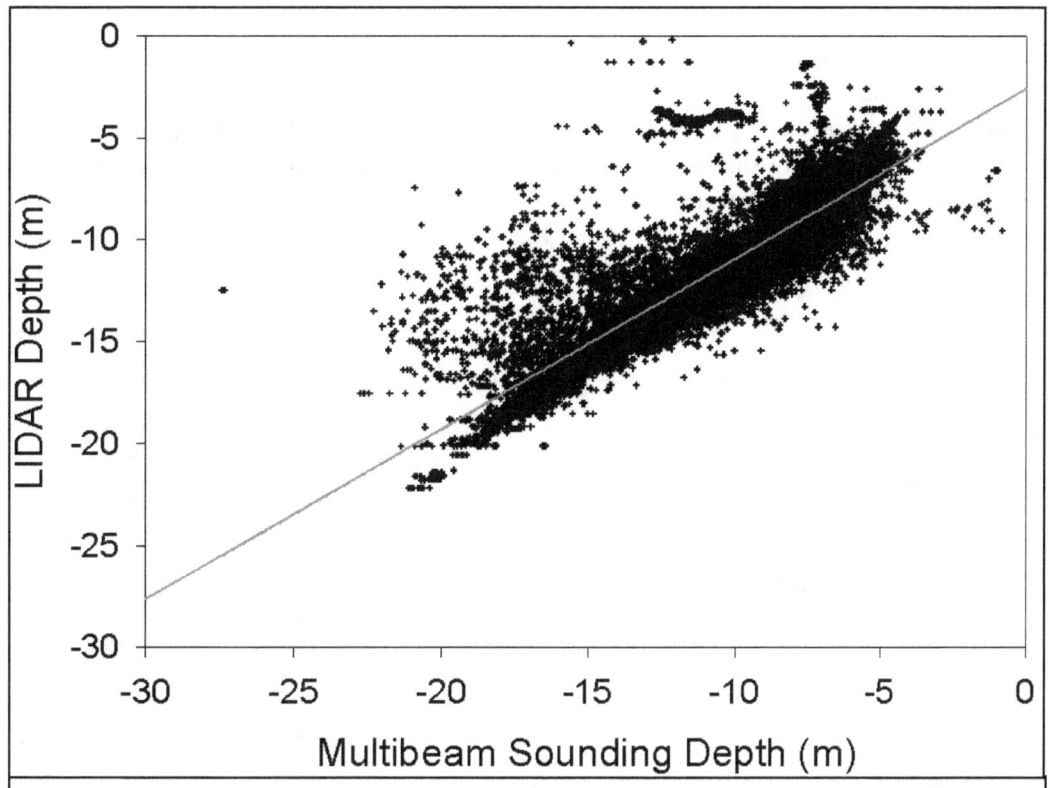

Figure 10. Linear least squares data model between multibeam soundings and LIDAR depths (m). LIDAR = -2.576+0.835*Multibeam sounding depth, R-Squared = 0.682.

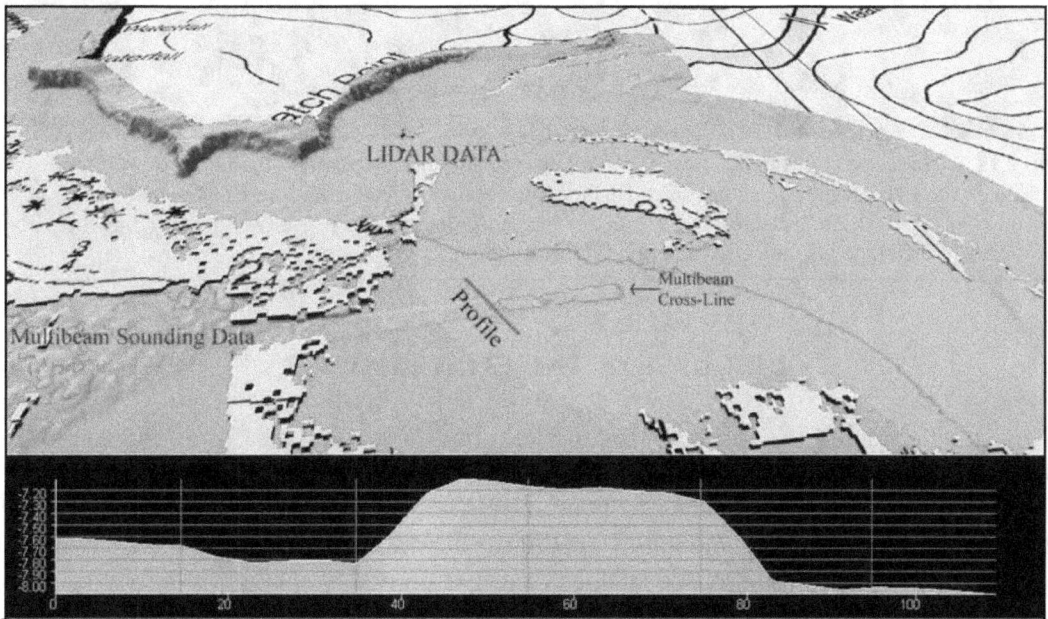

Figure 11. Multibeam and LIDAR xyz data merged into seamless DTM covering Makah Bay. Note the multibeam cross line is about 0.5 m more shoal than the LIDAR surface. Red line represents the location and extent of the bathymetry profile shown on the bottom. Purple contour is the 7 m bathymetry curve.

the cross line of multibeam data. The largest of the depth differences, those values in the extreme tails of the histogram (red portion of the sample range in Figure 8), are traceable to two distinct areas and appear to be artifacts of errors in LIDAR data cleaning due to interference from false bottom detection (Figure 12). The remaining general surface difference trend could be attributed to the fact that the multibeam data are cleaned and gridded with a shoal biased constraint to be more conservative with respect to dangers to navigation.

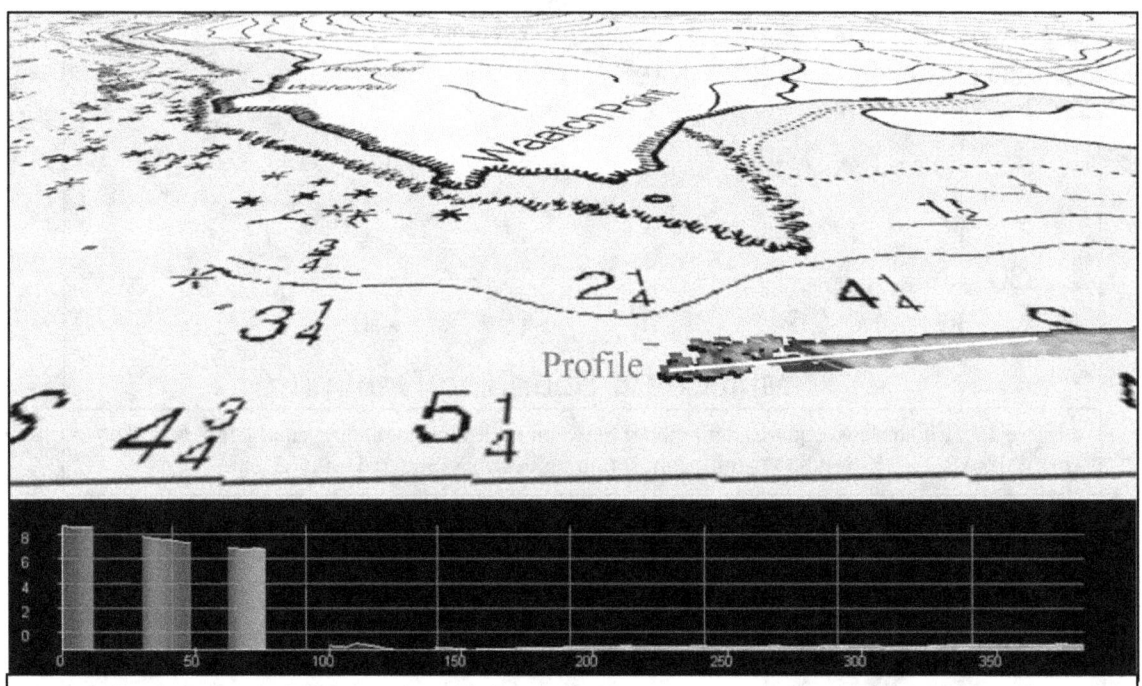

Figure 12. An area of significant surface difference (red and orange pixels) attributed to improper cleaning of the LIDAR data resultant from surface interference. The white line shows extent and location of profile, which indicates an 8 meter surface difference at the western end of the line and sub-meter surface difference (green and blue pixels) throughout the remainder.

DISCUSSION AND CONCLUSIONS

As with most coastal ALB surveys, floating kelp beds, breaking surf lines and turbid water were challenges to this survey. The window of opportunity for capturing favorable environmental conditions to successfully acquire ALB in coastal areas, such as OCNMS, is narrow. Avoiding the detrimental effects of turbid water created from excessively aerated breaking surf and increased surface runoff from heavy spring rains require data acquisition to be targeted for late spring to early summer. However, kelp growth and increased biological productivity can further impede the rate of acquisition success in this region beginning as early as May. Because of all these factors for attempting to acquire the LIDAR data for this particular survey, late April was selected.

The less than 2 meter swell and minimal wind-induced wave action that were experienced did not appear to negatively impact data quality over the course of the survey, indicating that ALB can successfully be achieved in this area at these particular sea states. Although sea state was favorable for this time of the year, recent heavy rainfall and a persistent low-lying fog layer severely reduced the productivity of this survey. Additional project funds would have been helpful to keep the mobilized flight crew on site while survey conditions continued to improve in order to maximize acquisition success.

The existence of a completed VDatum model covering this geographic region permitted the LIDAR data to be seamlessly merged with shallow water multibeam data using the same control datum. Of note, since OCNMS uses UTM coordinates to position all multibeam sounding data, it was required that the contractor deliver the NAVD88 referenced z-values in that same coordinate system. Unfortunately, it was later determined that VDatum Version 1.06 only accepts geographic coordinates, thus the positional information delivered by FPI in UTM had to be unprojected back to latitude and longitude (decimal degrees) prior to transforming the vertical units. This was a very time consuming step, which in retrospect, could have been avoided by simply not requiring the contractor to project the data prior to transformation. The VDatum tool would benefit from a future enhancement to address this limitation.

Despite relatively poor survey productivity resulting from poor flying conditions, a significant portion of nearshore bathymetry was acquired along the Cape Flattery coastline. The entire intertidal zone from Koitlah Point to Cape Alava, including numerous offshore rocks and islands, was also better defined by the high-resolution topographic LIDAR elevation data. The bathymetry data that were successfully acquired could potentially provide useful information for delineating rock features in many areas that were too hazardous to access through ship-based acoustic survey methods. Due to a lack of project funds and a desire to cover as much area as possible, this survey was designed to acquire 4 by 4 m spot spacing bathymetry data with just 100 percent coverage. Even though NOAA's OCS often requires 200 percent coverage for bathymetric LIDAR acquisition at a 4 by 4 spot spacing, it is possible that the data could still be used to update the Cape Flattery nautical chart 18485. Many of the least depths recorded on that chart are rather dated with some soundings potentially being obtained as early as 1834 (NOAA 2002). But it is important to stress that this survey was not designed or performed for nautical charting purposes and several critical levels of quality control were not implemented to evaluate and process the data for such purposes. Therefore the data will be submitted to NOAA's Hydrographic Survey Division for outside-source evaluation and archival at the very least. Various other elements within the NOS Hydrographic Surveys Specifications and Deliverables (2003) will also need to be assessed prior to electing the data for possible inclusion on the nautical charts.

LIDAR technology continues to evolve at a rapid pace. In the short time since this data was acquired, significant improvements have been made to the acquisition algorithms in order to increase extraction of more useful information for aiding environmental assessment (Francis and Tuell 2005). For example, in addition to acquiring depth and digital still frames, as with the SHOALS 1000T system used here, the new Compact Hydrographic Airborne Rapid Total Survey (CHARTS) system now has the ability to measure hyperspectral data, and it can produce seafloor reflectance images, as well (Tuell et al. in press). This new system could be an extremely useful tool for the National Marine Sanctuary Program (NMSP) by providing the ability to rapidly gain knowledge of both coastal and nearshore topography, combined with water column characteristics and seafloor reflectance – all of which are important components for understanding changes to the nearshore coastal and benthic environment.

ACKNOWLEDGEMENTS

The author would like to thank LCDR Richard Fletcher for assistance with obtaining the secchi readings, Dr. Richard Bouchard for providing the archived NDBC wave statistics, and Kathy Dalton for manuscript edits. The paper further benefited from comments provided by Dave Sinson, Pacific Hydrographic Branch, Seattle, WA.

REFERENCES

Francis, K. and G. Tuell. 2005. Rapid environmental assessment: The next advancement in airborne bathymetric LIDAR. Ocean News and Technology. May/June 2005.

Intelmann, S.S., J. Beaudoin, and G.R. Cochrane. 2006. Normalization and characterization of multibeam backscatter: Koitlah Point to Point of the Arches, Olympic Coast National Marine Sanctuary - Survey HMPR-115-2004-03. Marine Sanctuaries Conservation Series MSD-06-03. U.S. Department of Commerce, National Oceanic and Atmospheric Administration, Marine Sanctuaries Division, Silver Spring, MD. 22pp.

MacDonald, A. 2005. New developments increase use of airborne LIDAR bathymetry. SeaTechnology. September 2005. 46-48.

Milbert, D.G. 2002: Documentation for VDatum and a Datum tutorial. Vertical datum transformation software, Version 1.06. 23 pp.

MOA. 1998. Memorandum of Agreement between the U.S. Army Engineers (USACE) and the Naval Meteorology and Oceanography Command (COMNAVMETOCCOM) to establish the Joint Airborne LIDAR Bathymetry Technical Center of Expertise.

MOA. 2002. A Memorandum of Understanding Among the U.S. Army Engineer South Atlantic Division and the U.S. Army Engineer Research and Development Center and the Naval Meteorology and Oceanography Command and the National Oceanic Administration through the National Ocean Service and the Office of Marine and Aviation Operations to establish the Joint Airborne LIDAR Bathymetry Technical Center of Expertise. MOA-2002-047. 6pp.

National Research Council. 2004. A Geospatial Framework for the Coastal Zone, National Needs for Coastal Mapping and Charting. Committee on National Needs for Coastal Mapping and Charting. National Research Council of the National Academies. The National Academies Press. Washington, D.C. 168 pp.

NOAA. 2003. NOS Hydrographic Surveys Specifications and Deliverables. March 2003. 150pp. http://nauticalcharts.noaa.gov/hsd/specs/specs.htm.

NOAA. 2002. Nautical chart 15485. 15th Edition. December 2002. Marine Chart Division N/CS2, National Ocean Service, NOAA, Silver Spring, Maryland.

Riley, J.L. 1995. Evaluating SHOALS bathymetry using NOAA hydrographic survey data. 24th Annual Joint Meeting of UJNR Sea-Bottom Survey Panel. Nov 13-17, 1995, Tokyo, Japan. 10pp.

Spargo, E. A., Hess, K. W., White S. A., 2006. VDatum for the San Juan Islands and

Strait of Juan de Fuca with Updates for Puget Sound: Tidal Datum Modeling and Population of the Grids. U.S. Department of Commerce, National Oceanic and Atmospheric Administration, Silver Spring, Maryland, NOAA Technical Report, NOS CS 23. (In review).

Tuell, G. J.Y. Park, J. Aitken, V. Ramnath, V. Feygels, G. Guenther, and Y. Kopilevich. SHOALS-enabled 3-d benthic mapping, Proc. SPIE Vol. 5806. Algorithms and Technologies for Multispectral, Hyperspectral, and Ultraspectral Imagery XI. S. Chen and P. Lewis Eds. In press.

APPENDIX

Hydrographic & Topographic LIDAR Acquisition, Northwest Coast, Washington, Neah Bay to Cape Alava, WA Survey Report.

ONMS CONSERVATION SERIES PUBLICATIONS

To date, the following reports have been published in the Marine Sanctuaries Conservation Series. All publications are available on the Office of National Marine Sanctuaries website (http://www.sanctuaries.noaa.gov/).

Conservation Science in NOAA's National Marine Sanctuaries: Description and Recent Accomplishments (ONMS-06-04)

Normalization and Characterization of Multibeam Backscatter: Koitlah Point to Point of the Arches, Olympic Coast National Marine Sanctuary (ONMS-06-03)

Developing Alternatives for Optimal Representation of Seafloor Habitats and Associated Communities in Stellwagen Bank National Marine Sanctuary (ONMS-06-02)

Benthic Habitat Mapping in the Olympic Coast National Marine Sanctuary (ONMS-06-01)

Channel Islands Deep Water Monitoring Plan Development Workshop Report (ONMS-05-05)

Movement of yellowtail snapper (*Ocyurus chrysurus* Block 1790) and black grouper (*Mycteroperca bonaci* Poey 1860) in the northern Florida Keys National Marine Sanctuary as determined by acoustic telemetry (MSD-05-4)

The Impacts of Coastal Protection Structures in California's Monterey Bay National Marine Sanctuary (MSD-05-3)

An annotated bibliography of diet studies of fish of the southeast United States and Gray's Reef National Marine Sanctuary (MSD-05-2)

Noise Levels and Sources in the Stellwagen Bank National Marine Sanctuary and the St. Lawrence River Estuary (MSD-05-1)

Biogeographic Analysis of the Tortugas Ecological Reserve (MSD-04-1)

A Review of the Ecological Effectiveness of Subtidal Marine Reserves in Central California (MSD-04-2, MSD-04-3)

Pre-Construction Coral Survey of the M/V Wellwood Grounding Site (MSD-03-1)

Olympic Coast National Marine Sanctuary: Proceedings of the 1998 Research Workshop, Seattle, Washington (MSD-01-04)

Workshop on Marine Mammal Research & Monitoring in the National Marine Sanctuaries (MSD-01-03)

A Review of Marine Zones in the Monterey Bay National Marine Sanctuary (MSD-01-2)

Distribution and Sighting Frequency of Reef Fishes in the Florida Keys National Marine Sanctuary (MSD-01-1)

Flower Garden Banks National Marine Sanctuary: A Rapid Assessment of Coral, Fish, and Algae Using the AGRRA Protocol (MSD-00-3)

The Economic Contribution of Whale watching to Regional Economies: Perspectives From Two National Marine Sanctuaries (MSD-00-2)

Olympic Coast National Marine Sanctuary Area to be Avoided Education and Monitoring Program (MSD-00-1)

Multi-species and Multi-interest Management: an Ecosystem Approach to Market Squid (*Loligo opalescens*) Harvest in California (MSD-99-1)